HYDROLOGIC SCIENCE PRIORITIES FOR THE U.S. GLOBAL CHANGE RESEARCH PROGRAM

AN INITIAL ASSESSMENT

Committee on Hydrologic Science
Water Science and Technology Board
Board on Atmospheric Sciences and Climate
Commission on Geosciences, Environment, and Resources
National Research Council

NATIONAL ACADEMY PRESS
Washington, DC

NOTICE: The project that is the subject of this report was approved by the Governing Board of the National Research Council, whose members are drawn from the councils of the National Academy of Sciences, the National Academy of Engineering, and the Institute of Medicine. The members of the committee responsible for the report were chosen for their special competences and with regard for appropriate balance.

Support for this project was provided by the Army Research Office, National Aeronautics and Space Administration, National Oceanic and Atmospheric Administration, National Science Foundation (under Grant No. EAR-9814582), National Weather Service, and U.S. Geological Survey. Any opinions, findings, and conclusions or recommendations expressed in this material are those of the authors and do not necessarily reflect the views of the sponsors.

International Standard Book Number 0-309-06648-4

Additional copies of this report are available from:

National Academy Press
2101 Constitution Ave., NW
Box 285
Washington, DC 20055
800-624-6242
202-334-3313 (in the Washington metropolitan area)
http://www.nap.edu

Cover by Van Nguyen, National Academy Press. Adapted from painting *Niagara Falls* by Frederick Church, 1857.

Printed in the United States of America

COMMITTEE ON HYDROLOGIC SCIENCE

DARA ENTEKHABI, *Chair*, Massachusetts Institute of Technology, Cambridge
MARY P. ANDERSON, University of Wisconsin, Madison
RONI AVISSAR, Rutgers University, New Brunswick, New Jersey
ROGER C. BALES, University of Arizona, Tucson
EVILLE GORHAM, University of Minnesota, St. Paul
MARC B. PARLANGE, Johns Hopkins University, Baltimore, Maryland
CHRISTA PETERS-LIDARD, Georgia Institute of Technology, Atlanta
KENNETH W. POTTER, University of Wisconsin, Madison
ERIC F. WOOD, Princeton University, Princeton, New Jersey

National Research Council Staff

STEPHEN D. PARKER, Project Study Director
PETER A. SCHULTZ, Staff Officer
ANITA A. HALL, Project Assistant

BOARD ON ATMOSPHERIC SCIENCES AND CLIMATE

ERIC J. BARRON (*Co-Chair*), Pennsylvania State University, University Park
JAMES R. MAHONEY (*Co-Chair*), IT Group, Inc., Washington, D.C.
SUSAN K. AVERY, Cooperative Institute for Research in Environmental Sciences, University of Colorado, Boulder
LANCE F. BOSART, State University of New York, Albany
MARVIN A. GELLER, State University of New York, Stony Brook
CHARLES E. KOLB, Aerodyne Research, Inc., Billerica, Massachusetts
ROGER A. PIELKE, JR., National Center for Atmospheric Research, Boulder, Colorado
ROBERT T. RYAN, WRC-TV, Washington, D.C.
MARK R. SCHOEBERL, NASA Goddard Space Flight Center, Greenbelt, Maryland
JOANNE SIMPSON, NASA Goddard Space Flight Center, Greenbelt, Maryland
NIEN DAK SZE, Atmospheric and Environmental Research, Inc., Cambridge, Massachusetts
ROBERT A. WELLER, Woods Hole Oceanographic Institution, Woods Hole, Massachusetts
ERIC F. WOOD, Princeton University, Princeton, New Jersey

Ex Officio Members

DONALD BURKE, Johns Hopkins University, Baltimore, Maryland
DARA ENTEKHABI, Massachusetts Institute of Technology, Cambridge
EUGENE RASMUSSON, University of Maryland, College Park
JOHN ROADS, University of California at San Diego/Scripps Institution of Oceanography, La Jolla
PAUL WINE, Georgia Institute of Technology, Atlanta

NRC Staff

ELBERT W. (JOE) FRIDAY, JR., Director
LAURIE S. GELLER, Program Officer
PETER A. SCHULTZ, Program Officer
DIANE L. GUSTAFSON, Administrative Assistant
ROBIN MORRIS, Financial Associate
TENECIA A. BROWN, Senior Program Assistant
CARTER W. FORD, Project Assistant

The National Academy of Sciences is a private, nonprofit, self-perpetuating society of distinguished scholars engaged in scientific and engineering research, dedicated to the furtherance of science and technology and to their use for the general welfare. Upon the authority of the charter granted to it by the Congress in 1863, the Academy has a mandate that requires it to advise the federal government on scientific and technical matters. Dr. Bruce Alberts is president of the National Academy of Sciences.

The National Academy of Engineering was established in 1964, under the charter of the National Academy of Sciences, as a parallel organization of outstanding engineers. It is autonomous in its administration and in the selection of its members, sharing with the National Academy of Sciences the responsibility for advising the federal government. The National Academy of Engineering also sponsors engineering programs aimed at meeting national needs, encourages education and research, and recognizes the superior achievements of engineers. Dr. William A. Wulf is president of the National Academy of Engineering.

The Institute of Medicine was established in 1970 by the National Academy of Sciences to secure the services of eminent members of appropriate professions in the examination of policy matters pertaining to the health of the public. The Institute acts under the responsibility given to the National Academy of Sciences by its congressional charter to be an adviser to the federal government and, upon its own initiative, to identify issues of medical care, research, and education. Dr. Kenneth I. Shine is president of the Institute of Medicine.

The National Research Council was organized by the National Academy of Sciences in 1916 to associate the broad community of science and technology with the Academy's purposes of furthering knowledge and advising the federal government. Functioning in accordance with general policies determined by the Academy, the Council has become the principal operating agency of both the National Academy of Sciences and the National Academy of Engineering in providing services to the government, the public, and the scientific and engineering communities. The Council is administered jointly by both Academies and the Institute of Medicine. Dr. Bruce Alberts and Dr. William A. Wulf are chairman and vice-chairman, respectively, of the National Research Council.

Preface

The availability of fresh water is potentially one of the most pervasive crises of the coming century. Water-related decisions will determine the future of major ecosystems, the health of regional economies, and the political stability of nations. A vigorous program of research in hydrologic sciences can provide the basis for sound water management at local, regional, national, and international levels.

The Committee on Hydrologic Science was established by the National Research Council in 1999 to identify priorities for hydrologic science that will ensure its vitality as a scientific discipline in service of societal needs. This charge will be performed principally through a series of studies that provide scientific advice on the hydrologic aspects of national programs and U.S. hydrologic contributions to international programs.

This first report contains a preliminary assessment of the hydrologic science content of the U.S. Global Change Research Program (USGCRP). Because this is a short and focused report, little effort is spent to reaffirm the established and successful elements of the USGCRP. In fact, the Committee generally endorses the findings of the National Research Council (NRC) report *Global Environmental Change: Research Pathways for the Next Decade* (NRC, 1998a; the so-called *Pathways* report) in this respect. **Instead the attention here is directed toward the most critical missing hydrologic science elements in the FY2000 USGCRP. This brings the focus to the terrestrial component of the water cycle. The integrative nature of terrestrial hydrology could significantly strengthen the USGCRP**. Two specific examples of the useful roles of terrestrial hydrology are: (1) linking regional hydrologic and water resources systems with large-scale and global water and energy cycles and (2) coupling water and biogeochemical cycles through ecosystems. This report recommends science priorities on these and related topics.

This report was produced in a short period of time. The Committee first met February 8–9, 1999, at which time it several briefings from federal officials and scientists. The Committee met again April 6–8, 1999, and drafted report chapters at that time. Subsequently, Committee members edited and circulated materials until the report was completed. It is anticipated that in the next few years, several aspects of hydrology not dealt with in depth in this "initial assessment" will be fleshed out more thoroughly by the Committee and reported in more detail.

The Committee was aided in the study process by numerous agency liaisons, including L. Douglas James, National Science Foundation; Robert Hirsch, U.S. Geological Survey; Richard Lawford, National Oceanic and Atmospheric Administration; John Schaake, National Weather Service; Russell Harmon, Army Research Office; David Goodrich, U.S. Global Change Research Program Office; and representatives from the National Aeronautics and Space Administration. Three NRC staff members helped the Committee: Stephen Parker, director of the Water Science and Technology Board (WSTB), who served as principal staff officer for the Committee; Peter Schultz, staff officer with the Board on Atmospheric Sciences and Climate; and Anita Hall, a WSTB administrative assistant.

This report has been reviewed by individuals chosen for their diverse perspectives and technical expertise, in accordance with procedures approved by the NRC's Report Review Committee. The purpose of this independent review is to provide candid and critical comments that will assist the NRC in making the published report as sound as possible and to ensure that the report meets institutional standards for objectivity, evidence, and responsiveness to the study charge. The review comments and draft manuscripts remain confidential to protect the integrity of the deliberative process. We wish to thank the following individuals for their participation in the review of this report: Eric J. Barron, Pennsylvania State University; Stephen Burges, University of Washington; George M. Hornberger, University of Virginia; Dennis Lettenmaier, University of Washington; Diane M. McKnight, University of Colorado; Sharon E. Nicholson, Florida State University; Fred M. Phillips, New Mexico Institute of Mining and Technology; David H. Rind, NASA Goddard Institute for Space Studies; and Soroosh Sorooshian, University of Arizona. Although the individuals listed above provided constructive comments and suggestions, it must be emphasized that responsibility for the final content of this report rests with the authoring committee and the NRC.

It is the hope of the Committee that the recommendations are persuasive and that they will be pursued, as commitment to implementation of these recommendations should bring advances to hydrologic science for the benefit of society.

Dara Entekhabi, *Chair*
Committee on Hydrologic Science

Contents

Executive Summary

The availability of water to sustain life and to fuel economies is perhaps the most important recurrent constraint in human history, and it will remain so for the foreseeable future. During the past decade, an in-depth understanding of the water cycle, especially at regional scales, has emerged as a major scientific challenge within the U.S. Global Change Research Program (USGCRP), a federal effort to enhance understanding of the global environment and assess its possible evolution. As water is a critical component of other systems, it has emerged as a cross-cutting theme in the USGCRP. The global water cycle, now one of USGCRP's six fundamental program elements, offers two primary research challenges: (1) land-surface interactions and (2) atmospheric processes. Research in hydrologic science is primarily in the first area, an area that includes land surface–atmospheric coupling over a range of spatial and temporal scales and includes the role of the land surface state in climate variability and change. These challenges are important but limited. Broader challenges for hydrologic sciences that address cross-disciplinary research and recognize the integrative nature of terrestrial hydrology could strengthen the USGCRP.

Terrestrial hydrologic processes, specifically the storage and movement of water on land and within the terrestrial biosphere, are important across all of the USGCRP elements and should serve as a unifying physical process within the USGCRP. To meet these additional challenges, this report identifies two broad science areas that augment the current hydrologic sciences content of the USGCRP: (1) predictability and variability of regional and global water cycles and (2) coupling of hydrologic systems and ecosystems through biogeochemical cycles.

Predictability directly addresses the USGCRP priority of identifying possible future environmental change. This report recognizes current plans within the climate variability element but recommends additional research topics that can strengthen the long-term research goals of USGCRP. These additional topics include enhanced understanding of linkages in variability of global and regional hydrologic systems as the basis for producing improved predictions. The emphasis on variability and predictability, particularly in regional hydrologic systems, is designed to link the understanding of the global water cycle with emerging regional and local water resources issues.

Cross-disciplinary research involving hydrologic science is key to addressing challenges identified under both the USGCRP global carbon cycle and global water cycle elements. For example, terrestrial ecosystems exert a strong influence on the global water cycle through evaporation processes. Also ecosystem disturbances are likely to be a major pathway for any changes and shifts in water and chemical cycles resulting from human activity. The foundation for this research must be a better understanding of the water and chemical pathways and of hydrologic–ecosystem linkages and a new means of achieving this understanding. It is then possible to address the combined influences of climate change and land use change, which occur in the context of natural variability, on hydrologic systems and ecosystems.

The USGCRP should give high priority to developing effective measurement and data strategies specifically for the terrestrial component of the global water cycle. The strategies should address multiple needs, ranging from the detection of change to process studies to operational applications. Future planning for remote sensing and ground-based measurement networks should be integrated to give measurement strategies that are responsive to the priorities discussed above. This will require a high degree of interagency and international collaboration, and it will require new approaches to planning hydrologic measurements. Considerable attention also needs to be given to recovering and archiving hydrologic data and making the data available through effective data and information systems. These strategies need to integrate remote sensing and ground-based data, and they must be sustained over the long term.

Water issues are central to the USGCRP emphasis on global change and its impacts. Therefore water issues can help guide the evolution of new initiatives within the USGCRP. To yield effective results, concerted efforts need to be made to improve connections between hydrologic research and its applications.

1

Setting Priorities

Study of the water cycle is a priority in global environmental research because this cycle is central to the working of the Earth system. Furthermore, the availability of unpolluted water to sustain life and fuel economies is perhaps the most important recurrent constraint in human history, and it will remain so for the foreseeable future. During this past decade, concomitant with consideration and implementation of several of the findings in *Opportunities in the Hydrologic Sciences* (NRC, 1991), an in-depth understanding of the water cycle, especially at regional scales, has emerged as a major scientific challenge. This is reflected in the research priorities highlighted in several National Research Council (NRC) reports dealing with global change, climate variability, environmental quality, and hazards mitigation (NRC, 1998b–e). Moreover, along with carbon, water is one of the two major themes in the NRC report *Global Environmental Change: Research Pathways for the Next Decade* (NRC, 1998a; the so-called *Pathways* report), and water now has been established as one of the key program elements of the U.S. Global Change Research Program (USGCRP, 1999). The NRC and USGCRP reports generally present water as a critical component of other systems. From this perspective, many but not all priorities for hydrologic sciences emerge as components in cross-cutting programs such as the USGCRP. The *Pathways* report and the USGCRP implementation plan, *Our Changing Planet* (USGCRP, 1999), identify some key hydrologic science challenges, but there are also critical hydrologic issues that are not contained therein.

The USGCRP has progressed in its recognition of hydrology as a key issue. In the FY1999 edition of *Our Changing Planet* (USGCRP, 1998), hydrologic research and applications are discussed only in the context of climate variability and change research on seasonal-to-centennial time scales. In the FY2000 edition (USGCRP, 1999), however, the hydrologic cycle is identified as one of USGCRP's six fundamental program elements: (1) understanding Earth's climate system, (2) biology and biogeochemistry of ecosystems, (3) composition and chemistry of the atmosphere, (4) paleoenvironment/paleoclimate, (5) human dimensions of global change, and (6) the global water cycle. Terrestrial hydrologic processes, specifically the storage and movement of water on and within land and within the terrestrial biosphere, are important across all these elements and should serve as a unifying physical process of USGCRP activities.

The global water cycle element contains two primary research challenges. The first challenge relates to land surface interactions and the need to develop a better understanding of (1) the coupling of land surface hydrologic processes to atmospheric processes over a range of spatial and temporal scales; (2) the role of the land surface in climate variability and climate extremes; and (3) the role of the land surface in climate change and terrestrial productivity. The second relates to atmospheric processes and the need to develop a better understanding of (1) the role of clouds and their influence in the coupling of the atmospheric water and energy cycles and (2) the vertical transport and mixing of water vapor on scales ranging from the local boundary layer to regional weather systems.

The global water cycle program element bridges the gap in the spatial-scale spectrum between large-scale atmospheric research and smaller-scale hydrologic research. Its proposed endeavors that are related to the "fast component" of the climate system should simultaneously draw upon and feed into international programs such as the Global Energy and Water Cycle Experiment (GEWEX) and the Biological Aspects of the Hydrologic Cycle (BAHC). This program element also addresses important "slow component" aspects of the climate system, therefore relating it strongly to the Program on Climate Variability and Predictability (CLIVAR) as well.

These challenges are important, but limited. Broader challenges for hydrologic sciences that address cross-disciplinary research and recognize the integrative nature of terrestrial hydrology would strengthen the USGCRP. Two strategic research areas are identified in this report: (1) predictability and variability of regional and global water cycles and (2) coupling of hydrologic systems and ecosystems through biogeochemical cycles, including the characterization of water and chemical flow pathways at the surface–atmosphere and the surface–subsurface interfaces.

Within the first topic, predictability directly addresses the USGCRP priority of identifying possible future environmental changes and of defining the limits of prediction. The emphasis on variability, particularly in regional hydrologic systems, is designed to link understanding of the global water cycle with the emerging regional and local issues that are receiving increasing emphasis in USGCRP. Variability (and memory) in the global water cycle and regional hydrologic systems is due to both the cycling of water between reservoirs with various storage capacities (e.g., atmosphere, surface waters, near-surface soil moisture, and groundwater systems) and the development of feedback dynamics resulting from linkages among the reservoirs. Presently, for example, linkages to the groundwater reservoir are not considered in the USGCRP. The ultimate aim is to use the enhanced understanding of linkages in the variability of global and regional systems as the basis for producing improved and useful hydrologic predictions.

The second topic addresses priorities identified in both *Our Changing Planet* (USGCRP, 1999) and the *Pathways* report and highlights the need to understand linkages between the cycles of water, energy, nutrients, and carbon through ecosystems. Terrestrial ecosystems exert a strong influence on the water cycle through evaporation processes. Evaporative flux linking the surface and atmospheric systems and the recharge flux linking the surface and subsurface systems are two key components of the hydrologic cycle that are poorly understood and are very poorly monitored. It is imperative that the characterization of these two fluxes be recognized as grand challenges for hydrologic science.

Finally, ecosystem disturbances are likely to be a major pathway for any changes and shifts in water and chemical cycles resulting from human activity. This cross-discipline research area is key to addressing challenges identified under both the USGCRP global carbon cycle (contained within the *biology and biogeochemistry of ecosystems* USGCRP program element) and the global water cycle element. The foundation for this research must be a better understanding of water and chemical pathway, and of hydrologic system–ecosystem linkage, and a new means of achieving that understanding. It will then be possible to address the combined influences of climate change and land use change, which occur both in the context of natural and human-induced variability, on hydrologic systems and ecosystems.

2

Science Foundations and Basic Processes

PREDICTABILITY AND VARIABILITY OF REGIONAL AND GLOBAL WATER CYCLES

The cycling of water in its various states—solid, liquid, and gaseous—is a primary process within the Earth's climate system. Information on variability of states and fluxes over time is crucial for the understanding of the sustainability of local, regional, national, and international economies and ecosystems. It is essential to establish rates of cycling, changes in these resulting from human intervention, and the consequences of those changes for regional water availability. It is important to estimate the magnitude of potential changes in water reservoirs at the land surface (e.g., lakes, seasonal snow-packs, soil moisture, groundwater, glaciers, and ice sheets), changes in fluxes of water (e.g., precipitation, evaporation, runoff, and groundwater recharge), and changes in atmospheric water storage and transport, all of which have profound influences on the Earth's energy cycle and global change processes. Toward this end, a better understanding is needed of what causes both short-term and long-term variability in these fluxes that couple the land surface reservoirs of water with each other as well as with the oceans and atmosphere. It is a priority scientific objective to establish how much of the variability in the water cycle is predictable over a range of time and space scales.

Large-scale seasonal-to-interannual oscillations in climate have been shown to contribute significantly to the total variability in precipitation and temperature in some regions. These predictable patterns of variability create excellent opportunities for improving long-lead hydrologic forecasts based on measurements of climate indicators such as sea-surface temperature or snow-cover patterns.

An understanding of mechanisms linking large-scale climate variability with regional conditions also forms the basis for reducing the uncertainty associated with assessing regional impacts of global change over decadal-to-centennial periods. A region-specific ability to project the consequences of global change is now required, for example, by decision-makers concerned with long-term fixed capital investments in infrastructure such as dams, water diversion systems, and flood damage mitigation systems that are vulnerable to shifts in hydroclimatic regime. It is also required by policy makers debating whether possible shifts in hydroclimatic regime warrant increased measures to reduce greenhouse gas emissions (USGCRP, 1999).

The continental hydrologic cycle is the link between climate forcing at the large scale and surface impacts at the regional scale. Understanding how variabilities in regional hydrologic cycles are linked to large-scale climate is a research priority in both the hydrologic and climate sciences (NRC, 1998b).

Key challenges for hydrologic science are to define spatial and temporal regimes of hydrologic systems in which predictability is high and to characterize and understand the nature of variability in hydrologic systems. These two challenges are interrelated.

Predictability is the extent to which the future state of a system can be estimated based upon the (theoretical) availability of a comprehensive set of observations characterizing the system's initial condition. Once a hydrologic system has been shown to possess a substantial degree of predictability, useful prediction schemes can be devised to estimate the response of the system to external forcing (e.g., land use changes) or to variability of the climate system. The estimation of this response requires an in-depth understanding of (1) the relative contributions of local and remote forcing mechanisms to the total variability in hydrologic systems, (2) the changes in the response and the characteristics of hydrologic systems as they are monitored or modeled at different spatial and temporal resolutions, and (3) the nature of local and regional feedback mechanisms that affect the response of hydrologic systems to local and external forcing factors.

In this respect, the ability to make useful predictions—e.g., weather forecasting at a local scale (about 1–10 km) for hazards mitigation, preparation of water supply outlooks at the basin scale, and water availability projections under global change at a regional scale (about 10^2 to 10^3 km)—requires that the cycling pathways, the storage of water, and the nature of variability for physical processes be measured and understood.

Predictable Patterns of Seasonal-to-Interannual Variability

Issues of predictability form the essence of the scientific challenges associated with seasonal-to-interannual climate variability. The land, biosphere, atmosphere, and oceans are coupled together in an Earth system that has a wide range of time and space scales in its variability. For example, there are "slow" (e.g., deep groundwater and oceans) and "fast" (e.g., atmospheric water vapor and surface moisture) components in this system, whose rate of reaction is controlled in part by reservoir size, by the intrinsic rates of the processes involved, and by interactions with other processes.

The superposed variations with a variety of time scales means that there is potentially some predictability if the "slow" components can be isolated and monitored. For example, coupled air–sea interactions across the large tropical Pacific basin result in large-scale changes in climate that appear with some regularity (quasiperiodic) on a 2- to 6-year time scale. This phenomenon, known as the El Niño-Southern Oscillation (ENSO), has strong implications for hydroclimatic anomalies across the tropics, and it also affects conditions in the extratropics (NRC, 1998b). If the ENSO signal is linked to regional precipitation and temperature, then seasons-ahead predictions of the phenomenon can provide a means of narrowing the uncertainty associated with long-lead hydrologic predictions. The implications of such predictability for managing water resources, agriculture, electric power, and other climate-sensitive sectors are far–reaching (see Box A). ENSO is just one example; there are other phenomena in the Earth system that cause variations on interannual and longer time scales. For example, feedback between the atmosphere and snow cover, land moisture, vegetation cover, and regional water bodies other than the Pacific also induces systematic oscillations and excursions in regional climate.

Three key scientific questions in hydrologic science related to variability in the global water cycle and its regional seasonal-to-interannual predictability are as follows:

- The effects of large-scale seasonal-to-interannual oscillations in climate caused by interactions between land, oceans, and the atmosphere are apparent in many hydrologic records. The physical processes and cause–effect relations, however, are not understood well enough to support the design and implementation of forecasts that are sufficiently robust to be relied upon for regional water-related decision-making. *To what extent is regional-scale hydrology predictable? What types and locations of measurements will most enhance predictions?*
- Predictability of regional hydrologic systems is limited to large-scale climate predictability if forcing is unidirectional, i.e., large-scale climate affects regional hydrologic systems. If there is two-way land–atmosphere coupling or modulation of climate by local hydrologic processes, then there is potentially enhanced predictability gained through coupled hydrologic modeling. *Across which regions and seasons can predictions of regional water cycling be enhanced by robust coupled land–atmosphere modeling?*
- Hydrologic extremes (e.g., droughts and floods) may be influenced by large-scale climate variability and local land–atmosphere exchanges in ways that are quite distinct from how those influences affect regional hydroclimate under near-normal conditions. Current hydrologic forecasts (e.g., water supply outlooks or flow forecasts) are better at predicting near-normal conditions than they are at predicting extremes. Yet in terms of economic and environmental impact, predicting extremes is of much greater regional importance. *What special physical and statistical features (e.g., process pathways, influences across scales) can be used to link large-scale climate and regional-scale hydrology in the case of extreme events, and how are these features different for the case of floods and the case of persistent droughts?*

Sources of Long-Term Variability

Variability in hydrologic records includes contributions from both natural variability and human-induced changes of the landscape and climate. There are at least two critical gaps that research must address to separate the total variability into natural and human-induced contributions. First, the causes and spatial patterns of natural variability from both measured and paleoclimate records need to be understood. For example, tree-ring records in the southwestern United States show evidence of prolonged droughts that greatly exceed the magnitude and duration of droughts in the measured record, principally during the past 100 years. The extent and severity of these events within the region and the link to large-scale climate changes are not known. Second, it needs to be ensured that in future decades, scientists have the long-term hydrologic measurements needed to detect changes and that they have the understanding to link those changes to alterations in the landscape and in the broader-scale climate system. For example, observations are currently lacking to detect a change in the mass of the Earth's largest freshwater reservoirs, the polar ice sheets, to within 25 percent of the annual accumulation (Houghton et al., 1996). Further, the amounts and spatial patterns of groundwater recharge to critical water supply aquifers in the United States and other parts of the world have never been measured, in part because of a lack of understanding of how to make accurate measurements of recharge rates (Simmers, 1988).

Two key questions in hydrologic science related to understanding long-term sources of variability are the following:

Box A
Use of Seasonal Forecasts

The Salt River Project (SRP) is the largest provider of water and electricity in Arizona, delivering over 300 billion gallons of water per year to over one million persons, to industry, and to 238,000 acres of irrigated land in the metropolitan Phoenix area. SRP's water sources include both local groundwater and surface water from the 13,000-square-mile watershed of the Salt River, which originates in the mountains of eastern Arizona, and the Verde River, which flows from the north. SRP makes commitments to water allocations 6–12-months in advance based on reservoir storage levels. Over the long term, groundwater makes up about 10 percent of SRP's total delivery, though it can provide up to 30 percent of the total demand in a dry year when runoff is low. Much of the surface runoff originates as winter snowfall, making accurate forecasting of winter precipitation the primary year-to-year water supply question facing SRP. In the 1998 water year, which was a strong El Niño year, SRP for the first time used the seasonal forecast of a wet winter to influence its reservoir operations. SRP lowered water levels in the fall of 1997, in anticipation of a wet winter, to provide the water that they committed to deliver and to reduce the chance of spilling excess water the following year. The result was a savings of about $1.4 million from the reduction in the costs of groundwater pumping. As forecast, 1998 was a very wet year in the Southwest. Had it been an average or dry year, SRP would have had to spend an additional $3million to $5 million in groundwater pumping costs to make up the shortfall. Owing to the success of the strong El Niño signal in predicting winter precipitation in the Southwest in 1998, SRP plans to continue to use seasonal forecasts in decision-making and is helping drive research agendas for improved forecast information.

- Accurate and long-term measurements, both for understanding past variability and detecting future changes, are critical. From a water-resources standpoint, the long-term events of greatest concern are sustained droughts and increases in the magnitude and frequency of floods. However, this issue goes well beyond water resources, concerning essentially all of the Earth's hydrologic systems. *What combination of remote and in situ observations and paleohydrologic records are required to identify shifts in regional and local hydrologic properties resulting from both natural and human-induced factors?*
- Shifts in regional hydroclimate are linked to large-scale patterns that change over long time scales. Understanding linkages between regional and large-scale processes is essential to interpreting the record of past variability and to using that record as an analog of the expected shifts in regional water availability under global change conditions. *Are there spatial patterns in the variability of the hydrologic record that may serve as reliable predictors of the impacts of global change?*

Linking Measurements and Understanding Across Scales

Each process contributing to a hydrologic response, such as evaporation, snowmelt, surface runoff, or subsurface flow, has its own set of characteristic spatial and temporal scales. For example, the spatial resolution at which atmospheric systems are modeled is from 100 to 10,000 times the scale of heterogeneities in land surface topography, vegetation, soil texture, or snow cover. Similarly, variables like temperature or soil moisture can change significantly in a matter of hours, whereas vegetation changes occur seasonally. Measurements and models need to be designed so that they can aggregate processes at disparate scales.

Besides the variability in the original processes, some new scales of variability emerge when processes are coupled together. The critical challenge for hydrologic science is to understand and describe the ways in which heterogeneous processes interact with one another at different scales to produce the variability found in hydrologic systems. That understanding must then be translated into enhanced predictions. Two key questions follow:

• Advances in measurement technologies allow topographic, soil, vegetation and snow properties, and other parameters to be specified at finer and finer scales in predictive models. There are computational limitations, however, on the ability of hydroclimatic models to incorporate all processes and scales. Thus, a key question is to what extent the spatial structure and magnitude of the (fine-scale) variability in a particular parameter or variable is required for realistic modeling at the large scale (upscaling). The answer to this question depends on the scale at which a given process or variable interacts with other processes. For example, a prediction of snowmelt, runoff, and groundwater recharge over days to weeks may possibly benefit from a much finer description of soil and vegetation properties than would land surface feedback into a climate model for seasonal-to-interannual forecasts. For land–atmosphere modeling where the atmospheric boundary layer is the link in the exchanges, there is a natural spatial integration associated with turbulent mixing. *At what scales and for which processes should the spatial structure of surface heterogeneity be incorporated into the upscaling strategy for hydrologic models?*

• The coupling between the land, biosphere, and atmosphere imposes strong constraints on some hydrologic processes in some instances. For example, the diurnal cycle of surface moisture and energy flux at local scale is strongly affected by entrainment processes associated with the growth and collapse of the atmospheric boundary layer. Similarly, the seasonal cycle of these same fluxes at the regional scale is largely determined by the radiative and energy balance of the overlying atmosphere. *Of the physical constraints that come about because of the coupling of water and energy cycles, which may be used to bound the estimates of local and regional hydrologic fluxes?*

COUPLING OF HYDROLOGIC SYSTEMS AND ECOSYSTEMS THROUGH CHEMICAL CYCLES

The supply of water and its distribution over the landscape are primary determinants of nutrient inputs to ecosystems and to ecosystem productivity. Evaporation of water through ecosystems is a major control on the terrestrial branch of the hydrologic cycle. Cycles of water, energy, nutrients, and carbon have been identified consistently as priority cross-cutting themes in understanding environmental change (NRC, 1998e; USGCRP, 1999). Scientists currently lack the detailed knowledge of these coupled cycles required to make rational choices regarding the global issue of greenhouse gas emissions, and they lack detailed knowledge of the local to regional issues associated with human environmental disturbances such as land-use changes and their consequent perturbations to water quality and flow. To make the necessary environmental choices, which will have tremendous socioeconomic implications, a fundamental understanding is needed of how water exerts a controlling influence on ecosystems, and vice versa, with special emphasis on areas where human activity is having the greatest impact.

Characterization of Water and Chemical Pathways

Ecosystems and humans are affected by and rely on discharges of water from groundwater and surface water systems. These waters carry chemicals, including nutrients needed for survival and contaminants such as organic solvents or pesticides. Adverse effects include the possibility of chemical accumulation (e.g., selenium enrichment and increases in basin-scale salinity). Changes in climate or land use that affect groundwater recharge can cause changes in groundwater storage and discharge that could have adverse effects on humans and ecosystems. *To protect human welfare and the integrity of the ecosystems on which humans depend, understanding is needed concerning the pathways that water and chemicals follow as they move across the landscape and through the subsurface, as well as concerning the physical, chemical, and biological transformations that occur along these paths.*

Research has demonstrated that water and chemicals are transported most rapidly along preferential pathways (see Box B). The location and configuration of these pathways are influenced by the arrangement of soils and geological units having high permeability, as well as by cracks and fractures. Delineation of pathways is complicated by difficulties involved in characterizing the land surface (e.g., mapping vegetation and microtopography) and the inability to see into the subsurface. Pathway patterns can be postulated through the use of numerical models that track imaginary particles, which represent water molecules and/or chemicals. Results of such modeling studies allow scientists to quantify not only the particle pathways themselves, but also travel times and fluxes. The accuracy of these models depends on the characterization of subsurface media and the specification of boundary conditions (e.g., hydrologic fluxes).

Understanding interactions and pathways of water and chemical exchange between surface and subsurface hydrologic systems is impeded because there are many gaps in knowledge (e.g., measurement and understanding of key flux variables such as evaporation and groundwater recharge). If human welfare and the integrity of the ecosystems on which humans depend are to be protected, it will be necessary to address a number of priority science questions including the following:

- Lack of knowledge of the detailed distribution of subsurface materials is impeding the ability to track and predict the movement of solutes, including contaminants. Subsurface imaging has been an important tool for oil exploration for many years, and it has been used in groundwater investigations as well, but improved techniques are needed for addressing problems in both the petroleum industry and for hydrogeological site characterization. *Can geophysical techniques (e.g., ground-penetrating radar and nuclear magnetic resonance) be refined to provide ways of imaging the subsurface to provide information on the distribution of geologic units and preferential flow paths in a variety of complex geological settings?*
- Groundwater recharge to the water table and groundwater discharge to oceans, lakes, wetlands, and rivers are important to ecosystems and to human survival. Yet scientists are still struggling to measure the spatial and temporal distribution of groundwater recharge and discharge. *Can a general methodology be developed (e.g., using innovations in chemical, isotopic, and thermal measurements) to measure or otherwise estimate the spatial and temporal distribution of groundwater recharge and discharge (and fluxes of associated chemicals) over a basin? Can preferential paths of water and chemicals through the vadose zone be incorporated into basin-scale recharge theories?*
- Addressing large-scale problems such as basin-scale recharge, subsurface flow paths, or salinity sources will require an understanding of a number of fundamental issues that have been largely neglected over the past two decades. It also will require combining several powerful approaches that have largely

Box B
The Need for Basic Research in Subsurface Hydrology
Preferential Flow Paths for Water and Solutes

One of the most basic problems in subsurface hydrology concerns the pathway and rate at which water flows through porous and fractured media. This question is of obvious importance for irrigated agriculture, natural ecosystems, groundwater recharge, groundwater extraction, contaminant transport, and local, regional, and global water balances. The theoretical foundations for understanding and the mathematical descriptions were formed several decades ago, and incremental improvements have been made since then. Much of the research effort in this area over the past three decades has been linked with regulatory agency concerns over contaminant fate and, although this effort has supported the relatively short-term agency needs for decision-making, basic research in fluid flow has received much less emphasis. An example helps to illustrate the critical need for basic research that has broad implications for hydrology.

Water moves through the vadose, or unsaturated, zone lying between the ground surface and the water table as a wetting front that depends largely on properties of the media and on how fast the water is applied to the surface. The wetting front, however, does not move uniformly. Small perturbations along the air–water interfaces in soils and rocks give rise to fingers that move faster than the background wetting front. The result is that water and contaminants move and reach the water table faster and at higher concentrations than predicted based on current models. Whereas factors such as heterogeneity and layering are important, fingering also occurs in relatively uniform media. Once established, fingers in some media can elongate with time. Instability and fingering are of concern in the petroleum industry and limit the efficiency of secondary oil recovery, which involves injection of a fluid to help drive oil toward production wells. Although much has been learned from past work, basic research is needed to provide a foundation for applications as diverse as basin-scale recharge, salinity movement from irrigated areas, and isolation of wastes from water supplies.

Movement of contaminants in groundwater is controlled by preferential flow paths formed by connected geological units of high permeability, such as a buried river channel. It is exceedingly difficult, however, to find or predict where these high-permeability units occur in the subsurface. Consequently, efforts to track and predict the movement of contaminants in the subsurface are severely impeded. Also, contaminants that move into low-permeability units can be very difficult to remove, complicating efforts to clean contaminated sites (NRC, 1994).

been used independently in the past. These include (1) traditional geologic mapping and conceptual methodologies, (2) subsurface imaging, (3) numerical modeling, and (4) innovations in tracer techniques. *What combinations of methods will enable identifying the important flow paths for water and solutes in the vadose zone, and in the saturated zone, at scales from smaller than a hillslope up to regional aquifers?*

Interactions Between Hydrologic Systems and Ecosystems

In aquatic ecosystems, residence times, nutrient fluxes, and many other factors are determined by the rate at which water moves through the hydrologic cycle, yet little is known about how most terrestrial and freshwater ecosystems and the plants and animals that make up the ecosystems respond to changes in the hydrologic cycle. Being able to predict with confidence responses to climate variability and to anthropogenic perturbations requires great improvement in *understanding the interactions and feedback mechanisms between hydrologic systems and ecosystems.* Several aspects of these interactions deserve special attention:

- The next level of improvement in estimating regional evaporation from plants is likely to result from a combination of new observations (e.g., remote sensing) and models of vegetation that include photosynthesis and the assimilation of nutrients. *Can coupling between the cycling of nitrogen—a limiting factor for growth in many diverse aquatic and terrestrial ecosystems—and the cycling of water and carbon in models of land surface–atmosphere interactions improve the ability to estimate regional fluxes of water in terrestrial ecosystems?*
- The water-ecosystem linkage may be strongest but least understood in riparian systems. Riparian systems are under great stress because of human activities. These systems harbor a large majority of the regional biodiversity. Together with their associated groundwater systems, riparian regions also sustain human habitation and agriculture. Furthermore, these terrain features convey a large fraction of the exchange between the subsurface, surface, and atmosphere. Reliable tools for managing riparian communities do not exist, largely because of uncertainties in (1) riparian plant–water–nutrient relations, (2) basin boundary conditions, (3) physical hydrologic processes, such as riparian evaporation, over large areas, and (4) hydrological flow paths and residence times. *Can advances in basic knowledge of macro/micronutrient constraints vs. other physical/climatic constraints (e.g., water, energy) as controls on ecosystems, and especially new approaches for understanding and describing groundwater–surface water interactions and associated processes, provide the necessary understanding of riparian areas?*
- Long-term resource management of aquatic ecosystems must be based on a better fundamental understanding of how biogeochemical processes respond to the combined effects of climate variability, e.g., changes in the flux of organic matter, acid deposition, mineral weathering and of changes in the hydrologic cycle related to climate warming. Lakes and streams are largely buffered against changes in acidity by mineral weathering, the rate of which depends on the supply of acids from both precipitation and a basin's terrestrial ecosystem. *Can sufficient knowledge of biogeochemical cycles be developed (e.g., knowledge of long-term mineral-weathering rates and of coupled water, carbon, and nitrogen cycling) at small catchment to larger basin scales to enable scientists to make confident, multidecadal forecasts of basin and ecosystem response to perturbations, particularly where future anthropogenic effects may perturb sensitive headwater catchments beyond what they have previously experienced?*

Human Disturbances of Hydrologic Systems and Ecosystems

Human-induced perturbations now range from local to global in scale and may cause shifts that affect humans and ecosystems well beyond the source of the perturbation. Hydrologic shifts appear as changes in atmospheric water, precipitation, runoff, evaporation, groundwater recharge, or chemical cycles. *A major challenge facing scientists lies in separating the effects of human influences from climate variability in the hydrologic record.* Five issues of special concern regarding the hydrologic cycle, nutrient cycles, and ecosystems follow. The first two stem largely from global environmental issues; the latter three are related to land-use changes.

• Climate warming in response to increases in greenhouse gas concentrations is expected to accelerate the rate at which water moves through the global water cycle, with far-reaching influences on global ecosystems and economies. It is essential to establish current rates and possible changes in precipitation, evaporation, runoff, recharge, and atmospheric water vapor transports. Any assessment of climate change and of its causes and impacts must be based on significantly better observations of the water cycle. *What are the regional changes in the rate of water cycling and, with consequent changes in associated nutrient and contaminant cycles, how are the structure and function of the world's ecosystems influenced?*

• Knowledge of the rate of sequestration of carbon in lakes, reservoirs, peatlands (see Box C) and oceans, and knowledge of controls on those rates, is essential if the United States is to intelligently undertake measures to mitigate greenhouse gas emissions. The total amount of carbon in the combined pools, and the relative balance among these pools, is likely to be strongly influenced by global warming and particularly by changes in the water cycle. *How strongly will delivery of carbon to freshwater ecosystems and export of carbon from them be affected by changes in the hydrologic cycle, specifically the supply of water to these systems and in residence times within them ?*

• Contamination of rivers and lakes from point discharges are often traced to the source, effectively monitored, and strongly regulated. Contamination from nonpoint discharges remains poorly understood and is thus virtually unregulated. Although nonpoint source contamination occurs in both urban and rural environments, the latter presents the greater scientific challenge. Rural nonpoint source contamination is largely due to the erosion and transport of soil particles, to which nutrients (largely associated with fertilizers) and pesticides are attached. Riverine inputs of nutrients result in extensive zones of low oxygen concentration in coastal waters. Although the ability to describe soil erosion is relatively good, there has been very little success in predicting the transport of soil particles and associated contaminants over the land surface to streams and rivers. *What are the erosion rates, fluxes, and residence times for sediments transporting non-point-source contaminants and nutrients to downstream ecosystems?*

• Alteration of land use as a result of urban growth is inevitable, causing shifts in hydroclimate and generally increasing flooding and pollutant loading. There is also increased groundwater pumping and loss of groundwater recharge associated with urbanization. This decrease in groundwater recharge at least partially explains the degradation of aquatic systems in urbanizing areas. Understanding the nature of the dependence of various aquatic ecosystems on groundwater discharge is needed. Understanding is also needed concerning how to manage water in urban areas to maintain adequate groundwater flows to critical aquatic systems. *What are the changes in the hydrologic cycle, and in nutrient and contaminant cycles, caused by alterations in land use owing to urban land use (i.e., the conversion from natural landscape and agriculture to suburban developments)?*

• Wetland ecosystems are active in the cycling of water, nutrients, atmospheric trace gases, and carbon. Human activity has altered major wetland regions. Restoration of wetlands is seldom carried out based on a scientific approach. Wetland restoration is in particular need of attention, because in most situations there is little effort to establish, or to reconstruct, the prior condition, and follow-up is often absent or too brief for the degree of success to be evaluated. *Can the ability to use hydrologic science be improved substantially so that the damage to aquatic ecosystems caused by human disturbances can be repaired and, where possible, the ecosystems can be restored to their prior conditions?*

Box C:
Special Issues in High Latitudes

Ecosystems in high latitudes have been identified as being particularly sensitive to climate variability and warming. The critical roles of high latitudes in the global climate system range from freshwater inflow into the Arctic Ocean and its influence on thermohaline circulation to the modulating effects of cool temperatures, snow, and frozen ground on seasonal weather and hence on terrestrial ecosystem composition and production and on the storage or release of carbon. Northern landscapes contain vast stores of carbon capable of influencing—if released—the global climate system. For instance, the boreal and sub-Arctic forests of Canada (including some peatland forests) contain about 64 billion tons of carbon, whereas the peatlands of boreal and sub-Arctic Canada (some forested and some open) are thought to contain about 100 billion tons. These comprise a very considerable item in the carbon budget of North America. Much greater knowledge of the exchange of carbon among terrestrial, atmospheric, and oceanic systems within northern regions is necessary because human-induced changes in climate, fire frequency and intensity, and land use are likely to have profound effects upon it.

Both the storage of carbon in peat and the release of carbon by methane emissions are greatest when water tables are high. A lowering of water tables will result in a shift from anoxic to oxic conditions, shutting down methane emissions but releasing carbon dioxide owing to peat oxidation. In the southern boreal zone, climate warming may result in more frequent droughts and may cause fires that could smolder for years in remote peatlands, releasing large amounts of carbon dioxide and releasing small amounts of methane as a product of incomplete combustion. On the other hand, in the northern boreal and sub-Arctic zones, the melting of frozen ground is likely to flood and rejuvenate many peatlands, causing them to store carbon more effectively but also to emit larger amounts of methane. Understanding the relative importance of these two scenarios is critical to estimating their influence on the global carbon cycle and their feedbacks to the global climate system.

Recent studies of water, energy, nutrient, and carbon cycles in northern latitudes are sufficiently mature that a coordinated, integrated study to understand their coupling is possible. Results from such a study are expected to include (1) better understanding of the net ecosystem exchanges of carbon in boreal and tundra ecosystems, including seasonal, interannual, and decadal variability, (2) better estimates of freshwater inflow (and associated biogeochemical fluxes) into the Arctic, including its variability, and (3) a better understanding of the partitioning of terrestrial water and energy fluxes and their influence on both weather and climate at high latitudes.

3

Measurement and Data Strategies

In the history of hydrologic science, many of the significant advances have resulted from new measurements (NRC, 1991). Whereas there are often field experiments with specific research objectives, the hydrologic community has not placed a high priority on design and operation of observing systems that meet both operational and research needs. In addition, rarely has high priority been given to the establishment of long-term observing systems. Indeed, over the past decade, only modest progress has been made in hydrologic measurements, and data needs are much the same as delimited in *Opportunities in the Hydrologic Sciences* (NRC, 1991).

The science challenges identified in Chapter 2 have in common the need to characterize better the hydrologic cycle. The challenges fall into three categories. First, information is needed about the fluxes and storage of water as it moves through the hydrologic cycle and about associated fluxes of solutes, sediments, and energy. Second, hydrologic data are needed to monitor and understand changes in the quantity and quality of water. Third, in the traditional scientific sense, hydrologic data are needed to test hypotheses and models and to formulate new hypotheses. Data are needed at a variety of scales. However, the spatial and temporal scales of available data limit our ability to investigate hydrologic variability and to detect changes. For most fluxes, a fundamental impediment to progress is how to relate point measurements to fluctuations in hydrologic variables at large spatial scales, in heterogeneous terrain, and over long periods of time.

Existing data networks should be viewed as critical elements within the USGCRP that are needed to answer its identified research challenges. These networks should be viewed as opportunities upon which to build the data sets needed in the next century to address questions regarding global change. Resolution of the questions in Chapter 2 requires new data, with measurement location, frequency, duration, and accuracy characteristics that go beyond the capabilities of existing networks. Allocation of new resources for data collection and data archiving must seek a compromise between scientific and operational needs, between ground-based networks and remotely sensed measurements, and between information on water quantity and quality. In establishing new systems or revamping old ones, the needs of operational water management as well as those of research and long-term monitoring should be considered carefully. Such foresight can obviate costly retrofitting or attempts to analyze data trends from operational systems in which the data noise may be greater than the signal. The establishment of any data system must take into account cost-effectiveness, but should do so with a long-range view.

Maintaining and Upgrading Ground-Based Networks

Research in hydrologic science relies heavily on data from operational ground-based networks, which need to be upgraded to meet evolving needs. In some cases adding relatively low-cost sensors or expanding the coverage of existing networks could provide the measurements to help meet the scientific challenges identified in Chapter 2. For example, thermistors to measure soil and snow temperature, devices to estimate soil water content, and instruments to sense snow properties are readily available but are deployed only to a limited extent as part of operational networks. In other cases, even large financial investments for incremental expansion of instrument networks will leave important questions about hydrologic fluxes and reservoirs unanswered. Thus, technical and analytical innovations are necessary to overcome the paucity of useful hydrologic data being collected and collated.

The need for long-term measurements is becoming increasingly clear in investigations of environmental change, including changes to the hydrologic cycle. Despite the increasingly recognized importance of data records of long duration, only a few dedicated organizations have successfully maintained high-quality data-collection efforts over periods of 50–200 years. Research organizations have experienced difficulties in committing limited monies year after year to monitoring a task often not viewed as research. There are also problems associated with obtaining continuous and homogeneous observations from operational networks. These networks often designed for purposes other than long-term research tend to change in response to new technology, to changing operational needs, and to budget pressures. Detecting hydrologic change requires data sets of greater quality and reliability than are often needed for research on processes or for operational forecasting. Federal agencies that make up the USGCRP must demonstrate their commitment to the collection, archiving, and dissemination of data suitable for the hydrologic research challenges identified in the FY2000 *Our Changing Planet*. In particular, USGCRP and its member agencies should reevaluate existing ground-based networks and consider future networks in light of the critical need to detect hydrologic change in future decades.

Most hydrologic data have been collected with local-, regional-, or national-scale questions in mind. Some of the science issues in Chapter 2 are local or regional, but others are continental or global. Because it will not be possible with known technology and present-level resources to monitor hydrologic phenomena over the entire Earth, there is a need to actively promote international cooperation to implement networks that yield the maximum return.

Under the current budgets, there is a decreasing number of measurement points that federal, state, and local agencies can maintain, and there is a limit to the number of measurements that they can produce. For example, the national streamgaging network, maintained by the U.S. Geological Survey (USGS), is a unique and irreplaceable source of primary data supporting hydrologic research, planning, and management. It is of critical national importance that this source of long-term, consistent, and reliable data be maintained and upgraded to support the measurement of both streamflow and stream stage, especially in extreme conditions such as floods (USGS, 1998; NRC, 1999a). Federal agencies must implement effective measures to reverse the trend in river flow monitoring network losses. This can be achieved only through both inter- and intra-agency commitments to making long-term measurements in support of work on recognized national priorities in water resources planning, water quality management, aquatic ecosystem protection and restoration, hazard mitigation, and global change research. Implementation plans should address stable funding partnerships to maintain the networks.

Similarly, a national network of groundwater monitoring wells is essential for groundwater recharge characterization, an important scientific challenge presented in Chapter 2. These data are also central to the identification of long-term trends related to pumping, drought, and climate and land-use change. Although the cost of making water-level measurements is often relatively low (measurements are sometimes made by volunteers), the costs of coordination, data management, and well maintenance are relatively high. Current funding is insufficient to maintain the present network, with the result that some states have been forced to drop wells from their networks (e.g., 43 wells—25 percent of the total were dropped—from the network in Wisconsin owing to decreased funding). New approaches should be explored to ensure that a long-term network of monitoring wells for both water level and water chemistry is maintained, such as the establishment of new partnerships between the federal and state agencies that manage the network. Furthermore, funding sources should be stabilized to ensure that the network is maintained.

Integration of Remote Sensing with Ground-Based Measurements

Over the past decade the ability of remote sensing to provide spatial data on rainfall, soil moisture, snow cover, snow water equivalence, and other key hydrologic variables has been demonstrated. As recognized in FY 2000 *Our Changing Planet*, the capabilities of remote sensing are powerful, in part because remote sensing has the unique capability of providing consistent, global coverage. Remote sensing has the potential to improve our understanding of terrestrial hydrology. However, much of this capability remains underused in terrestrial hydrologic research and applications because of a lack of investment in research and technology by the USGCRP agencies. Important technological questions remain to be answered before data with sufficient accuracy and reliability can be collected and systematically applied for measuring and detecting changes in land surface hydrologic properties. Moreover, hydrologic science has yet to articulate, or in some cases even recognize, the necessary synergy between remotely sensed and ground-based measurements. Remote sensing will not preclude the need for ground-based measurements. However, it is not likely that existing ground-based networks are adequately configured to take maximum advantage of remote sensing measurements. Further research is required to determine the mix of remote sensing and ground-based networks to address several of the questions posed in Chapter 2.

The four areas that offer the greatest promise for satellite remote sensing to further hydrologic research and science are rainfall, soil moisture, snowpack properties, and surface-water level monitoring. Significant investment in remote sensing technology and hydrologic research is needed to prepare adequately for future missions in these areas. Two additional areas that have been initiated by other disciplines, but in which hydrologic science has an obvious interest, are vegetation properties and ice sheet mass balance.

Following the success of the Tropical Rainfall Monitoring Mission (TRMM) in measuring rainfall using a combination of active and passive microwave observations, the National Aeronautics and Space Administration (NASA) initiated planning for a global precipitation mission. Rainfall is associated with weather systems that display considerable spatial and temporal variability, requiring a sampling frequency of multiple times per day. The proposed global precipitation mission provides measurements that are central to the understanding of the predictability and variability of regional and global water cycles (Chapter 2). It also directly addresses USGCRP priorities in the global water cycle, particularly weather prediction and weather-climate relationships.

Space-based remote sensing offers the potential of measuring soil moisture, one of the main hydrologic variables that is not measured in any large spatial array of ground-based instruments. Soil moisture is the most significant indicator of the state of the terrestrial hydrologic system, and it is the governing parameter for partitioning rainwater among evaporation, infiltration, and runoff. It also provides a critical link between the physical climate and biogeochemical cycles. No existing or planned instrument meets these needs; thus, NASA has initiated studies for an exploratory mission. Soil moisture measurement is clearly a high priority within the global water cycle, and planning should be initiated for a measurement system that can effectively blend ground-based and satellite measurements.

With the launch of Terra, the hydrologic community will have the possibility of developing more accurate snow-covered-area products using the moderate resolution spectrometer (MODIS) instrument. The gain in accuracy comes from the improved spatial and spectral resolution of MODIS over the NOAA's Advanced Very High Resolution Radiometers (AVHRR) instruments. However, as with AVHRR, MODIS's surface observations will be hampered by clouds. An instrument with at least the resolution of MODIS should be made available to provide snow cover data for the foreseeable future. Research has shown that further advances in measuring snowpack properties from space will come only with a multispectral, multifrequency microwave sensor, either through active synthetic aperture radar (SAR) or through a high-resolution radiometer. Both systems offer the promise of measuring snow water equivalent, with a lower-frequency, higher-resolution SAR being preferred for measuring deep snow properties in mountainous regions, and with higher-frequency passive sensors being preferred for thinner snow packs on the Great Plains. A SAR sensor for snow water equivalence thus helps address perhaps the most significant water resources management challenge in the western United States—that of seasonal water supply forecasting. Over 85 percent of the Colorado River streamflow derives from snowmelt from mountainous regions, and late-spring water-supply forecasts based on estimates of how much snow is on the ground still differ from measured seasonal runoff by 20 percent to 50 percent in the various subbasins. SAR data have also been shown to be useful in estimating ice sheet dynamics. Better snow observations in the plains can help forecast floods like the record 1997 flood on the Red River of the North, and they can help in predicting moisture availability in agricultural areas on the Great Plains.

A fourth exploratory area that shows great potential is river discharge and changes in inland lake levels. Although river stage can be conveniently observed in situ for small to moderate rivers, discharge information for many of the world's major rivers is not available. Monitoring river discharge and lake storage would allow for the closure of the terrestrial water balance at continental scales and would provide important data on the linkage between the terrestrial and oceanic hydrospheres. Such monitoring would potentially yield an important indicator of global change. Precision altimetric sensors developed for oceanographic and ice-sheet measurements are potentially capable of detecting changes of a few centimeters in inland water surfaces.

In addition to these areas, continued work in quantifying and characterizing land cover changes is of critical interest to the hydrologic science community, as well as to the fields of ecological and biogeochemical research (NRC, 1999a). Terrestrial vegetation is an important determinant of hydrologic conditions through its effects on the surface radiation balance, on precipitation–runoff partitioning, and on evaporation. Remotely sensed data provide the only means by which variables such as growth form (e.g., woody vs. herbaceous vegetation), leaf type, and growing season onset and termination can be estimated at high resolution. Research and applications—currently using AVHRR, Landsat, and SPOT (Satellite Pour l'Observation de la Terre), and by extention MODIS—need to be extended to include data from the Vegetation Canopy

Lidar (VCL) mission. The use of VCL data for biological and hydrologic monitoring, process studies, and model parameter estimation needs to be expanded, since this mission will provide unique data on vegetation structure that may identify vegetation stress related to climatic factors and landscape (or land use) change. As discussed subsequently, multivariate model assimilation schemes might be used to incorporate these and other in situ ecosystem observations with other in situ and remotely sensed hydrometeorological data sources into a consistent framework. Such assimilations would facilitate subsequent hydrologic (and ecological) process studies, as well as provide a check on the underlying physics and parameterizations of the coupled ecological–hydrologic models used in the assimilation process.

Earth satellite missions, data management, and science to support development and use of data from these missions form the centerpiece of the USGCRP (e.g., NASA's Earth Science Enterprise). Potentially, this arrangement can serve hydrology well. However, for hydrologic science and applications to get the maximum benefit from the missions, a much closer link with ground-based network design, data management, and science support will be needed. Further, the ready availability of remote sensing information poses several specific challenges to the hydrologic community: (1) classical data-poor hydrologic models designed to function (and calibrate) with sparse in situ observations of precipitation and stream discharge need to be augmented and possibly replaced with new, process-resolving distributed hydrologic models that are forced with multispectral remotely sensed (and ground-based) observations, (2) basin-scale and regional validation databases consisting of coordinated in situ and remote-sensing data collection programs need to be established, and (3) hydrologists must adjust from being passive recipients of limited remote-sensing observations to acting as a unified scientific community engaged in supporting the definition, design, and implementation of satellite remote sensing missions. The USGCRP agencies with research and applications programs in hydrologic science must be proactively supportive of these goals.

Data Interpretation: Synergy in Modeling and Measurement

Advances in data interpretation capabilities are required to use the measurements effectively in the context of science questions and societal needs. Data assimilation (merging of models and data), solution techniques for inverse problems (inferring system properties from sparse data with the aid of models of the system), nested regional atmospheric models (and hydrologic models), and expanded analysis capabilities of geographical information systems (GIS) are opening new possibilities for the interpretation and use of data.

Data assimilation, an approach routinely used in meteorology and oceanography, is a particularly promising aid in the integration of large volumes of multisource and multiscale data as well as in the interpretation of measurements. It is an extension of standard model calibration. However, data assimilation is data-driven whereas calibration is model- and data-driven. Thus, one of the primary purposes of data assimilation is to produce data products that are directly useful for hydrologic analyses. For example, sequential remotely sensed data of ground temperature and microwave emissivity for the top few centimeters of soil may be used in conjunction with a soil column model to infer profiles of soil temperature and moisture down to tens of centimeters. In the context of groundwater, tracer and water level measurements in a sparse network of monitoring wells may be assimilated into a numerical groundwater model to infer soil hydraulic properties throughout the aquifer. Another use of data assimilation is to demonstrate the impact of under used and nontraditional data types (e.g., remote sensing, groundwater levels, and new in situ networks) on hydrologic predictions. The framework may be used to analyze the value of various observing systems through so-called "data-denial" experiments. In these experiments the value of a particular data source can be quantified in terms of the effect it has on estimates of a particular hydrologic variable. The design of new, cost-effective observing networks may be similarly done in the framework of data assimilation.

Nested regional climate models provide a bridge between the spatial scales of atmospheric and land surface processes. Understanding predictability in hydrologic systems and tapping that predictability for planning, assessments, and forecasts clearly require bridging tools that can make use of detailed spatial data to estimate spatial responses. In the western United States, capturing precipitation and runoff patterns requires topographic information of detail considerably greater than that found in even the most detailed general circulation models.

Advanced GIS tools make spatial modeling and interpretation possible. For example, assessing the impact of land-use changes on runoff or recharge requires information on the time evolution of land use as inputs to hydrologic models. GIS tools are likely to become a central part of hydrologic modeling.

Supporting Long-Term Experimental Sites

Long-term experimental sites for characterizing the water balance, flow pathways, and reactive transport provide the data essential for developing models and methods to scale hydrological variables, characterize basin-scale variability, and understand the limits of predictability. In the future they should also provide the data essential to understanding the coupling between hydrologic systems and ecosystems, and they should provide measurements for the calibration and validation of remote sensing data. The goal is to operate dozens of such basins worldwide over decades and across a range of bioclimatic zones, geological settings, and land-use characteristics. Existing programs in the United States (e.g., the National Science Foundation's Long-Term Ecological Research sites; the U.S. Department of Agriculture's experimental watersheds; the Water, Energy, and Biogeochemical Budget research watersheds of the USGS; and the AmeriFlux CO_2 sites partially supported by the U.S. Department of Energy and NASA) provide a useful starting point; some were designed as hydrological study sites, but others do not have a significant hydrological component. Agencies that support these sites should encourage multidisciplinary observations and studies. International and interagency coordination activities need to be launched to augment these sites for hydrologic research in a cost-effective manner. At larger scales, activities such as the USGS's National Water Quality Assessment (NAWQA) program (Box D) are needed to monitor and to understand the trends in the quality (and quantity) of the nation's surface and subsurface water resources for both science and policy purposes. Furthermore, more sites are needed in the United States and in selected climatic zones around the world in order to characterize hydrologic processes across the full range of environmental conditions.

Data Accessibility and Quality Control

Advances in hydrologic science will depend largely on how well investigators can integrate reliable, large-scale, long-term data sets. Inevitably, many scientists from a variety of disciplines and backgrounds will be involved in data collection and analysis over a significant period of time. Creating effective data

Box D
National Water Quality Assessment Program

The National Water Quality Assessment (NAWQA) program was implemented by the U.S. Geological Survey in 1991 to provide information that will help assess historic, current, and future water quality conditions in representative river basins and aquifers in the United States. A primary objective is to describe relations between natural factors, human activities, and water quality conditions and to define those factors that most affect water quality in different parts of the nation. The NAWQA design provides comparable information on water resources in 60 important study units (river basins and aquifers) across the nation. Together, these study units account for 60 percent to 70 percent of the nation's water use and population served by public water supplies and cover about half of the land area of the United States. The similar design of each investigation and use of standard methods make comparisons among the study units possible. From these unit studies, regional and national assessments can be made. These regional and national assessments, referred to as "national synthesis," aggregate water quality information and also analyze on a national basis important issues such as nonpoint source pollution, sedimentation, and acidification. Out of consideration for cost and management capacity, intensive assessment activities in each of the 60 NAWQA study units are being conducted on a rotational basis, with one-third of the units being studied intensively at any given time. For each study unit, 3- to 5-year periods of intensive data collection and analysis are alternated with 5- to 6-year periods of less intensive study and monitoring.

systems for assembling and distributing scientific data sets is not trivial. For example, better access to data from the Next Generation Weather Radar (NEXRAD) WSR-88D Doppler weather radar network with national coverage can significantly enhance the capability of monitoring precipitation fields for operational and research applications in the United States (see Box E). Many successful data systems have been constructed within the scientific and operational communities. Despite technical and political challenges, these data systems have high scientific value.

Many hydrologic data are currently not accessible and are in danger of being lost because of poor archiving by agencies or by responsible individuals as they retire or move to new assignments. Underuse of other hydrologic data stems from problems with quality control. For example, the limited use of ground-based radar rainfall data outside of the operational environment is partially attributed to the lack of research-quality data products and partially to poor archiving practices. Hence, resolving these issues would enhance the scientific basis and the routine use of these data in hydrology. In another example, downhole geophysical logs collected during the drilling of groundwater wells provide accurate continuous records of subsurface properties and are a valuable source of information. These records are not maintained in an electronic archive that is readily accessible and usable and they represent a tremendous missed opportunity to capture data that could have many and varied future uses. The logs, plus information regarding the location and accuracy of well logs from various sources, including commercial well drillers, need to be archived.

There is clearly a need for spatially distributed, regular information on water quality to address the priority science questions in the coupling of hydrologic systems and ecosystems through chemical cycles noted above. Although the long-term water quality monitoring efforts in the United States have been well designed to meet their own, often focused, goals, they are generally not connected, do not share data, and do

Box E
The Next Generation Weather Radar (NEXRAD)

The Next Generation Weather Radar (NEXRAD) WSR-88D Doppler weather radar network was created to meet the weather surveillance radar needs of the National Weather Service, the Federal Aviation Administration, and the Department of Defense. Its primary objective is to create nearly continuous radar coverage of the continental United States, making it possible to monitor the occurrence and development of severe weather events, including tornadoes, damaging winds, hail, heavy precipitation, flash floods, and hurricanes (NRC, 1995). Although the WSR-88D system has successfully improved forecasts and short-term warnings of severe weather events, the full promise of the system has not yet been realized because of (1) unreliable data archiving for subsequent research and analyses, (2) the lack of affordable accessibility to the data by the educational and research communities, and (3) the relative inaccuracy of precipitation estimates from the radar data. A recent report by the NRC's GEWEX panel (NRC, 1999b) recommends ways in which the utility of NEXRAD data might be improved for quantitative studies of the evolution of weather systems over long time periods and large spatial scales. If implemented, these strategies could help leverage the billion-dollar investment in NEXRAD into a capability that further enhances understanding of the precipitation processes associated with climate variations and flood-producing weather events.

Box F
Volunteer Measurements and K–12 Education

Cooperative data collected by volunteers who are not paid professionals has been successfully used in monitoring weather, groundwater levels, and water quality. However, the notion of involving students and the public directly in research and in the protection of hydrologic systems, specifically concerning water quality data, has left many traditional research scientists wondering, "What about quality assurance and control (QA/QC)?" A number of programs have been successful at developing a culture of QA/QC responsibility and serve as models of how students and citizens can be engaged in monitoring water quality and collecting high-quality baseline data on their local waterways.

Global Learning and Observations to Benefit the Environment (GLOBE) is a worldwide network of students, teachers, and scientists working together to study and understand the global environment. Students and teachers from over 6,500 schools in more than 80 countries are working with research scientists to learn more about our planet. GLOBE students make environmental observations (hydrology, meteorology, soils, and other measurements) at or near their schools and report their data through the Internet (www.globe.gov). Scientists use GLOBE data in their research and provide feedback to the students to enrich their science education. In addition to meeting science objectives, GLOBE science and education activities help students reach higher levels of achievement in science and math. GLOBE helps to increase the environmental awareness of all individuals while increasing our scientific understanding of the Earth. The Rivers of Colorado Waterwatch Program, nurtured by the Colorado Division of Wildlife, was started earlier and served as an inspiration for GLOBE. The Izaak Walton League's Save Our Streams program has also been recognized for its solid monitoring work. These and other parallel programs have developed multiyear records of stream water quality at locations where career scientists do not sample or only infrequently sample. Nationwide, over 340,000 volunteers are involved in over 500 water quality monitoring programs. These volunteer organizations can be strong partners in the nationwide monitoring strategy (ITFM, 1995).

not provide a broad, comprehensive assessment of water quality at continental, regional, or watershed scales. At present there is not an adequate, readily accessible archive for water quality data for the nation. The largest water quality archive in the United States is the Storage and Retrieval system (STORET), maintained by the U.S. Environmental Protection Agency. STORET incorporates data from other governmental agencies, from volunteers, and from any other programs or agencies that choose to submit data (see Box F). In 1997, STORET received data from 67,000 sites, 1,200 of which were from the USGS networks. As a long-term archive, STORET needs to be improved with respect to the documentation of its quality assurance and quality control procedures, and the more consistency of its reporting in order to reduce large spatial and temporal data gaps. Systematic criteria that are cost-effective for collecting and archiving valuable water quality data need to be defined and implemented as part of operational networks.

4
Applications and Knowledge Transfer

Water cycle issues affect all segments of society in all parts of the world. Everyone has a stake in the climate, in water resources, in the ecosystem, and in the economic implications of changes in the amount and distribution of fresh water on Earth. Scientific research in hydrology must have close links with societal and stakeholder needs, or applications; it must also have a vigorous knowledge-transfer effort that addresses kindergarten through university education, public outreach, and transfer of research results into practice.

Human population growth has resulted in heavy demands on the quantity and quality of water resources worldwide. The sustainability of these water resources in the twenty-first century will depend on the intelligent management of water resources systems under a more variable (and possibly warmer) future climate. The development of improved management strategies and viable interventions to meet these challenges will entail unprecedented coordination and integration across a broad range of research disciplines and segments of society.

Applications and User Integration

The increasing emphasis of the USGCRP on integrated assessment of the regional impacts of global change, including impacts on water resources, is fostering participatory, interactive research involving researchers, decision-makers, resource users, educators, and others who need more and better information about climate and its impacts. USGCRP is also supporting efforts to improve climate forecasting and the use of climate forecasts to manage water, fisheries, forests, crops, and range lands at the regional and local levels.

The term "regional assessment" is used to describe the collection, interpretation, evaluation, and communication of information of relevance to decision-makers, resource managers, and other individuals interested in a specific geographic location. Such assessments are "integrated" in several ways: (1) an end-to-end integration, which links understanding of ocean–atmosphere processes to regional impacts and policy responses, (2) an interdisciplinary integration, which brings together natural and social scientists, (3) the integration of scales of analysis from the local to the global and from short to long time scales, and (4) the integration of university and government research agendas with private sector concerns and public interests.

dialog with stakeholders has shown that in nearly all regions of the United States, there is a strong demand for additional climate information and particularly for how climate changes will affect water availability, quality, and demand. This dialog has also revealed critical research needs in hydrologic science that are necessary to meet this need. Increased attention to the research topics in Chapter 2 is supported by the assessment activities to date, and their is a clear need to continue this sort of formal assessment to guide the effectiveness and applications of hydrologic research.

Education and Knowledge Transfer

Greater hydrologic literacy facilitates transfer of hydrologic information from research to application. Promoting greater literacy includes educating decision-makers, the public and K–12 students in order to support informed public decisions about water issues that society will face in coming decades. New approaches in undergraduate and graduate education that transcend the traditional focus on hydrologic applications are needed to educate a generation of scientists and engineers who can develop and apply advances in hydrologic science (NRC, 1991).

The hydrologic science community should adopt a stronger sense of responsibility for delivering timely and relevant scientific tools (e.g., instruments, models, and procedures) to the operational community to meet societal needs. In turn the operational community should break the isolation that has caused lags in incorporating advances in hydrologic science in operational procedures. Effective two-way knowledge transfer between researchers and the many public agencies and private sector individuals who work with hydrology-related issues is clearly needed.

5
Conclusions and Recommendations

Hydrologic science is becoming increasingly recognized as an important element of global environmental research. *Our Changing Planet* (USGCRP, 1999) identifies several new key research priorities in hydrologic science that have not previously been well addressed. The following conclusions briefly assess the state of hydrologic science research within the USGCRP, and consider not only the aims of the USGCRP, but also the broader research community. These conclusions are supported by the discussions in chapters 2-4 of this report on science priorities, measurement and data strategies, and applications and knowledge-transfer needs. Recommendations corresponding to the conclusions are included.

Conclusion 1: The development of scientific capability to detect and predict changes to the water cycle in response to natural and human-induced climate variability is a key priority research area. The development of such scientific foundations requires strategic investments in both measurement and basic research programs. Current USGCRP water cycle activities have a strong emphasis on climate and on the influence of oceanic and atmospheric processes on climate. Several priority areas for research in hydrologic sciences are currently not well developed in the USGCRP.

Recommendation 1.1: *The identification of the limits to predictability of hydrologic variables should be among the guiding scientific challenges of the USGCRP water cycle initiative. The dependence of these limits on space and time scales, the sources of variability, and the effects of interactions among terrestrial, atmospheric, and oceanic components on variability need to be addressed in the context of predictability.*

Recommendation 1.2: *Agencies should establish mechanisms to foster multi-disciplinary research on hydrologic system–ecosystem coupling. The emphasis on ecosystems will require that aqueous chemical cycles become prominent research priorities of the USGCRP water cycle initiative; at present, water quality is essentially absent from the USGCRP.*

Recommendation 1.3: *Research on climate–hydrology linkages should be broadened to address issues of groundwater recharge, evaporation, basin-scale water balance, water and chemical pathways, and ecosystem responses. Fundamental research in these areas is also a priority in contexts other than climate, such as changing land use and its impacts on watershed-scale hydrologic processes.*

Recommendation 1.4: *Detailed implementation plans should be developed for the priority scientific challenges identified in this report. These plans should be based on a comprehensive view of current understanding, a futuristic view of technology, and a realistic view of cost-effective strategies. The agencies with substantial roles in hydrologic science research and measurements need to make certain that these efforts are led by stewards with effective authority and organizational support.*

Conclusion 2: The satellite measurement programs that constitute over 70 percent of the USGCRP water cycle budget in FY2000 are important but are not sufficiently focused to meet the specific challenges posed in *Our Changing Planet* or outlined in this report. Remote sensing is particularly well suited for global change research needs such as the characterization of precipitation, snow-pack properties, and surface soil. Realizing the full potential of remote sensing for hydrologic research and applications will, however, require a well-integrated satellite, ground network, and information management program. At present, the USGCRP gives a much lower priority to ground-based measurement networks and to long-term monitoring as critical elements in hydrologic research. Many hydrologic data archives are not readily accessible, and they are thus in danger of being lost.

Recommendation 2.1: *The USGCRP should give a high priority to developing effective measurement and data strategies for the terrestrial component of the water cycle. The design of new instrumentation and monitoring networks needs to incorporate effectiveness requirements for detecting change as well factors related to operational forecasting and process-level research. These networks need to integrate remote sensing and ground-based data, and they must be sustainable over the long term.*

Recommendation 2.2: *A study of data and measurement strategies for hydrologic science should be initiated immediately and should be completed within about a year. Considerable attention needs to be given to recovering and archiving hydrologic data and making it available through effective data and information management systems. The study should include a process for broad participation from the research and applications community in hydrologic science and in related sciences.*

Recommendation 2.3: *A parallel study should be initiated to assess the current state of and need for long-term experimental sites. Resources for expansion of these programs to include measurement and monitoring components required for hydrologic process studies and basin-scale water balance should be given high priority.*

Conclusion 3: Research in hydrologic science has much to contribute to the USGCRP emphasis on global change impacts, especially on water resource management issues (e.g., performance of water delivery systems, decision-making under drought conditions, or monitoring of the quality and quantity of regional groundwater and surface water systems). That emphasis also provides an important framework to help define future priorities in hydrologic science.

Recommendation 3.1: *Water resources management should be an integral, visible component of the global water cycle research initiative and should help guide the evolution of new initiatives within the USGCRP.*

Recommendation 3.2: *Formal efforts should be initiated to develop a better means of connecting hydrologic research and its applications in water resources management (i.e., two- way communication and knowledge transfer between researchers and the management community is needed).*

Recommendation 3.3: *New initiatives should be launched in hydrologic education and literacy that lead research scientists, the public, and government and other agencies to become stake holders of a common interest in the water resources of a changing world.*

References

Houghton, J. T. 1996. Climate Change 1995: The Science of Climate Change. Cambridge, Mass.: Cambridge University Press.

Intergovernmental Task Force on Monitoring (ITFM). 1995. The Strategy for Improving Water Quality Monitoring in the United States—Final Report of the Intergovernmental Task Force on Monitoring Water Quality. U.S. Environmental Protection Agency, Washington, D.C. and U.S. Geological Survey, Reston, Va.

National Research Council (NRC). 1991. Opportunities in the Hydrologic Sciences. Washington, D.C.: National Academy Press.

NRC. 1994. Alternatives for Ground Water Cleanup. Washington, D.C.: National Academy Press.

NRC. 1995. Assessment of NEXRAD Coverage and Associated Weather Services. Washington, D.C.: National Academy Press.

NRC. 1998a. Global Environmental Change Research: Pathways for the Next Decade. Washington, D.C.: National Academy Press.

NRC. 1998b. A Science Strategy for U.S. Participation in GOALS (Global Ocean-Atmosphere-Land System) Component of the CLIVAR (Climate Variability and Predictability) Programme. Washington, D.C.: National Academy Press.

NRC. 1998c. Decade-to-Century-Scale Climate Variability and Change: A Science Strategy. Washington, D.C.: National Academy Press.

NRC. 1998d. GCIP (Global Energy and Water Cycle Experiment (GEWEX) Continental-Scale International Project): A Review of Progress and Opportunities. Washington, D.C.: National Academy Press.

NRC. 1998a. Hydrologic Sciences: Taking Stock and Looking Ahead. Washington, D.C.: National Academy Press.

NRC. 1999a. Ecological Indicators for the Nation. Washington, D.C.: National Academy Press (in press).

NRC. 1999b. Enhancing Access to NEXRAD Data—A Critical National Resource. Washington, D.C.: National Academy Press.

NRC. 1999c. Hydrologic Hazards Science at the U.S. Geological Survey. Washington, D.C.: National Academy Press.

Simmers, J. 1988. Estimation of Natural Groundwater Recharge, NATO ASI Series C: V. 222. Boston, Mass.: D. Reidel Publishing Co.

U.S. Geological Survey. 1998. A New Evaluation of the USGS Streamgaging Network: A Report to Congress. Reston, Va.: U.S. Geological Survey.

U.S. Global Change Research Program (USGCRP). 1998. Our Changing Planet: FY1999 U.S. Global Change Research Program. Washington, D.C.: Office of Science and Technology Policy.

USGCRP. 1999. Our Changing Planet: FY2000 U.S. Global Change Research Program. Washington, D.C.: Office of Science and Technology Policy.

Appendix

Biographical Sketches of Committee on Hydrologic Science

Dara Entekhabi *(chairman)* is an associate professor in the Department of Civil and Environmental Engineering and the Department of Earth, Atmospheric and Planetary Sciences at the Massachusetts Institute of Technology. His research interests are in the basic understanding of coupled surface, subsurface, and atmospheric hydrologic systems that may form the bases for enhanced hydrologic predictability. Specifically, he conducts research in land–atmosphere interactions, remote sensing, physical hydrology, operational hydrology, hydrometeorology, groundwater–surface water interaction, and hillslope hydrology. He received his B.A and M.A. degrees from Clark University. Dr. Entekhabi received his Ph.D. in civil engineering from the Massachusetts Institute of Technology.

Mary P. Anderson is a professor in the Department of Geology and Geophysics at the University of Wisconsin, Madison. Her current research interests include the effects of potential global climate change on groundwater–lake systems and quantifying groundwater recharge. Dr. Anderson received a B.A. degree in geology from the State University of New York at Buffalo and a Ph.D. in hydrology from Stanford University. She is a former member of the Water Science and Technology Board.

Roni Avissar is professor and chair of the Department of Environmental Sciences and director of the Center for Environmental Prediction at Rutgers University. His research focuses on the study of land–atmosphere interactions from micro to global scales, including the development and use of a variety of atmospheric, land, and oceanic models. Dr. Avissar received his B.S. degree in soil and water science, his M.S. degree in micrometeorology, and his Ph.D. in mesoscale meteorology from the Hebrew University in Israel. He is the editor of the *Journal of Geophysical Research-Climate* and *Physics of the Atmosphere*.

Roger C. Bales is a professor in the Department of Hydrology and Water Resources at the University of Arizona. Dr. Bales conducts research on the hydrology and biogeochemistry of alpine areas, polar snow and ice, and water quality. He received his B.S. degree from Purdue University, his M.S. degree from the University of California, Berkeley, and his Ph.D. from the California Institute of Technology.

Eville Gorham is a professor of ecology and botany at the University of Minnesota. His research interests are in ecology and biogeochemistry, in particular ecosystem acidification (both natural and anthropogenic) and the role of northern peatlands in the global carbon cycle and their likely responses to climate warming. He received his B.Sc. and M.Sc. degrees from Dalhousie University in Nova Scotia and his Ph.D. from the University of London. Dr. Gorham is a member of the National Academy of Sciences and the Water Science and Technology Board, and he has served on several NRC committees.

Marc B. Parlange is a professor in the Department of Geography and Environmental Engineering at Johns Hopkins University. His primary research interest is in hydrology and fluid mechanics in the environment, especially questions of land–atmosphere interaction, turbulence and the atmospheric boundary layer, watershed-scale hydrology, and vadose zone transport processes. Dr. Parlange received his B.S. degree from Griffith University (Brisbane, Australia) and his M.S. degree in agricultural engineering and Ph.D. in civil and environmental engineering from Cornell University.

Christa Peters-Lidard is an assistant professor in the School of Civil and Environmental Engineering at the Georgia Institute of Technology. Her primary research focuses on measurement and modeling of terrestrial water and energy balances and fluxes related to land–atmosphere interactions over a range of temporal and spatial scales. This research encompasses the areas of micrometeorology, boundary layer meteorology, field experiments, hillslope hydrology, hydrometeorology, numerical modeling, spatial data analysis, and remote sensing. She received her B.S. degree in geophysics from Virginia Polytechnic Institute and State University and her M.A. and Ph.D. degrees in civil engineering and operations research from Princeton University.

Kenneth W. Potter is a professor of civil and environmental engineering at the University of Wisconsin, Madison. His teaching and research interests are in hydrology and water resources, including hydrologic modeling, estimation of hydrologic risk, estimation of hydrologic budgets, watershed monitoring and assessment, and hydrologic restoration. Dr. Potter is a past member of the Water Science and Technology Board and has served on many of its committees. He received his B.S. degrees in geology from Louisiana State University and his Ph.D. in geography and environmental engineering from The Johns Hopkins University.

Eric F. Wood is a professor in the Department of Civil and Environmental Engineering, Water Resources Program, at Princeton University. His areas of interest include hydroclimatology with an emphasis on land–atmospheric interaction, hydrologic impact of climate change, stochastic hydrology, hydrologic forecasting, and rainfall-runoff modeling. Dr. Wood is an associate editor for *Reviews in Geophysics, Applied Mathematics and Computation: Modeling the Environment,* and *Journal of Forecasting.* Dr. Wood received a Sc.D. in civil engineering from Massachusetts Institute of Technology. He is a member of the Water Science and Technology Board, the Board on Atmospheric Sciences and Climate (BASC), and BASC's Climate Research Committee.